Dedicated to all the young people
**who dare to ask questions
and follow their curiosity**
to the edges of the universe.

Written by: Cara Mayo. **Designed by:** Jacqui Fenner.
Copyright © 2017 by Jacqueline Fenner and Cara Mayo. Illustrations copyright © 2017 by Jacqueline Fenner.
SEEKERS OF SCIENCE SERIES and all related characters and elements are trademarks of Jacqueline Fenner and Cara Mayo.
All rights reserved. This book or any portion thereof may not be reproduced or used in any manner whatsoever without the
express written permission of the publisher except for the use of brief quotations in a book review.

First Printing, 2017

ISBN 978-1546396956

Published by: Cara Mayo and Jacqui Fenner
seekersofscience@gmail.com

Science Consultants: C. Alex Young (Associate Director for Science, Heliophysics Science Division, NASA Goddard Space
Flight Center), Steele Hill (SOHO Mission Media Specialist, Wyle Information Systems Inc.), and Louis Mayo (Astronomer/
Program Manager, ARIES Scientific Inc.).

Disclaimer: This is a work of fiction. While the science and historical references in this book are accurate, the names, characters, businesses, places, events and incidents are either the products of the author's imagination or used in a fictitious manner. Any resemblance to actual persons, living or dead, or actual events is purely coincidental.

 **// Seekers of Science Series**

Jordan and the Solar Eclipse

Written by **Cara Mayo** // Designed by **Jacqui Fenner**

Jordan rubbed her eyes wearily as she looked out the car window. The Sun was just starting to peak out over the horizon. She and her mom had been driving since early that morning. They were on their way to a NASA event to view the solar eclipse taking place that afternoon.

Jordan
Middle School Student

Jordan's mom was an amateur astronomer. She didn't have a degree in science, but she loved learning about the planets and stars just the same. Her favorite activity, since Jordan was little, had been to take her telescope out to the backyard on a clear night and look up at the night sky. When Jordan was younger, she had loved staying up with her mom and looking through the telescope at far away planets. Sometimes, they would race to see who could find a planet or constellation first. Other times, Jordan would use a book of planets, stars, and galaxies and challenge her mom to find them through her telescope. Or sometimes, they would simply lie down in sleeping bags and trace their own constellations through the sky. They would giggle as they explained the origins of their constellations. But over the years, Jordan spent fewer and fewer nights in the backyard with her mom. Her attention had moved from space and onto other, more important things — homework, soccer practice, and hanging out with friends.

Maria
Amateur Astronomer

Still, her mom's enthusiasm for outer space never wavered.

Jordan's mom had learned about the upcoming solar eclipse through a group for other amateur astronomers and decided to plan this mother-daughter outing. Jordan didn't understand why a solar eclipse was so exciting. Nor did she feel like she cared. None of her friends knew what a solar eclipse was and she felt a little nerdy telling them that she would be missing a day of school to see it. It was her mom who had dragged her out of bed saying she would regret missing this special experience. Jordan had reluctantly agreed.

Soon, they arrived at a large, grassy field dotted with people and their cars, telescopes, and tents. Jordan and her mom drove around until they found a small patch of grass of their own. Her mom jumped excitedly from the car. Jordan wasn't so quick to move — they had a whole day of this ahead of them...no need to rush right now. They had only been setting up for a few minutes when Jordan heard a voice yell over the crowd, "Maria!"

Jordan's mom turned around to see a tall, lanky, middle-aged man bounding across the field toward them, in a way that reminded Jordan of their family dog. He had a camera bag swinging at his side. "Hello, Hue!" shouted Jordan's mom.

"Maria, you're not going to believe the people I've been meeting," said Hue, "Astronomers, astrophysicists, heliophysicists, engineers...There's even a space historian named Parmesh whose stories you won't believe! Come on, I want to introduce you to someone." Hue paused, realizing that someone else was there, and looked over Maria's shoulder at Jordan. "Well! You must be Jordan." Jordan nodded shyly. Hue continued, oblivious to her shyness. "I've heard so much about you. You're so lucky to be out here. I'd bet most kids your age don't even know solar eclipses exist, and you get to see one firsthand! I wish my mom had taken me to one of these events when I was younger." He beamed at her and Jordan shrugged, not knowing what to say.

"Oh! I almost forgot, take these." Hue handed Jordan and her mom a pair of paper glasses with shiny, paper-thin lenses. "You're going to need them. Remember, you can't look at the Sun without these solar glasses until it's fully covered by the Moon."

"These flimsy glasses are going to protect our eyes...?" Jordan questioned, "They don't seem very powerful."

"They might not look powerful," Hue held the glasses up to his face, "but make no mistake, these glasses work just like the solar filters on telescopes."

Jordan still felt skeptical, but followed her mom's lead and put the glasses in her pocket. "Anyways, Maria, I want to introduce you to a friend. She's an astronomer at NASA. Jordan you should come along too!"

solar glasses

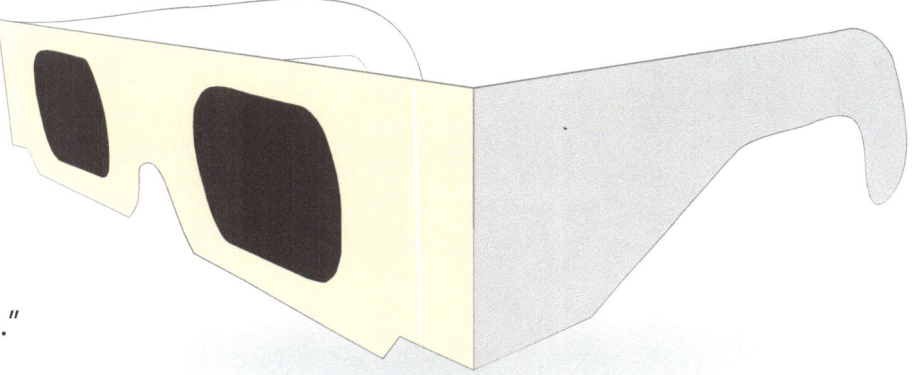

Hue
Amateur
Astronomer
& Photographer

Reluctantly, and realizing that there was no reason for her to say "No", Jordan followed her mom and Hue across the field to a large, green tent. There, she found a young lady sitting at a table on her laptop computer. Her vibrant, red hair was held together in a messy bun and she was peering with a focused intensity at the screen

through dark-rimmed glasses.

"Hi, Olivia" Hue said, tapping on the metal pole of the tent as if it were a door.

For a second, Olivia was so engrossed in what she was doing that she didn't even look up. Hue cleared his throat to get her attention. "Oh, I'm so sorry Hue, I was in the zone...Hi!" Olivia said.

"Olivia, I want you to meet Maria and her daughter, Jordan." Olivia reached out to shake their hands, "It's a pleasure to meet you." Jordan was struck by how young Olivia looked. Weren't scientists usually...older?

"So, Jordan, are you excited to be missing school today?" She winked and flashed a contagious smile. Jordan couldn't help but smile back bashfully, "I guess it's nice, but I still had to get up just as early." Olivia laughed, "It will be worth it, don't worry!"

"Yeah, my mom keeps telling me that," said Jordan, smirking up at her mom. Maria put her hands on her hips and smirked back as if to say, If only you'd listen to me sometimes...

"You know, I had parents just like yours." Olivia said to Jordan, as she played absent-mindedly with a moon pendant on her necklace. "They would drag me all over to see astronomical events in-person and I never really appreciated it back then. But look where I am now! I'm an astronomer working for NASA. I can't imagine doing anything else." She looked up thoughtfully. "Honestly, I don't think I would have even considered this field if it weren't for my parents. It might be hard to see why this is so exciting now, but I think down the road you'll be happy you came. Who knows, maybe you'll even change your mind by the end of the day," Olivia said with a wink.

Jordan considered this for a moment. It was doubtful, she thought, but didn't want to disappoint Olivia. So she shrugged and said, "I guess so."

"I could actually use your help on something, Jordan," Olivia said, gesturing Jordan over to the table. "Would it be okay if she helps me out, Maria?"

"Of course," Jordan's mom said with a smirk. She looked over at Hue and asked, "want to help me set up my equipment?" Hue nodded and, after agreeing that Jordan would meet her back at the car in an hour, Maria and Hue left the tent.

Jordan walked over to Olivia's table where several items were scattered about. There were pictures of the Sun, scissors, markers, and tape.

"Part of why I'm at this event is to put together an activity for students, like you, to teach them about solar eclipses. Tell me, what do you know about solar eclipses?"

Jordan thought for a moment. "Well, my mom told me that a solar eclipse is when the Moon comes right between the Sun and the Earth. And because of that, the Moon blocks out the Sun. Everything is supposed to get really dark, like it's nighttime during the day."

"Exactly." said Olivia, "Now, what if I told you that the Sun is 400 times bigger across than the Moon. How would the Moon be able to cover something that big?"

Jordan's eyes widened. "The Sun is that much bigger than the Moon?" Jordan asked incredulously. "I didn't know that...I'm not sure."

"It's pretty incredible, right?" Olivia stood up and taped one of her pictures of the Sun to the side of the tent. She asked Jordan to stand by her a few feet away from the paper Sun. "Give me a 'thumbs up' so that your thumb is covering at least part of our paper Sun." Jordan obeyed, and threw her other hand on her hip, striking a pose for good measure.

"Let's pretend that your thumb is the Moon and you are the Earth. This Sun is obviously much too big for your Moon to cover it completely. Now close one eye. Keep that eye on your thumb and take a few steps back. What do you see?"

As Jordan did this, she saw the Sun get smaller. Or was her thumb getting bigger? "My thumb is covering up more of the Sun!" Jordan exclaimed. "So, maybe if I stand all the way at the end of the tent, my thumb will cover the whole Sun." Jordan ran to the other end of the tent and put her thumb up – the Sun was gone. She smiled before she could stop herself.

"Very good!" said Olivia, "Now let's take it one step further. How do you think the Moon covers an entire Sun during a solar eclipse?"

size to scale

size
as perceived
from a distance

*fact: The Sun is 400X (400 times) bigger in diameter (across) than the Moon.

Earth   Moon   Sun

1X

??? X

distance
from Earth to Moon
during a solar eclipse

distance
from Moon to Sun
during a solar eclipse

*Diagram is not to scale and is for illustration purposes only.

Jordan needs your help

Activity | Try the Sun activity at home yourself. Based on what you've learned from that activity, how do you think our relatively tiny Moon covers the giant Sun during a solar eclipse? See the diagram on the left to help you figure it out.

Your Answer:

Olivia
NASA
Astronomer

Jordan thanked Olivia for showing her the activity and left the tent. Her spirits were just a little higher than they were before she had entered. As she walked, she thought about the size of the Sun compared to the Moon and the Earth and imagined how big the Sun would look if she were an astronaut in space.

A low moan caught Jordan's ear as she passed by a van. She stopped and looked around. A boy, about the same age as her, was looking at a poster titled, "Types of Solar Eclipses." Underneath the titles were three words – Total, Partial, and Annular. He had three pictures in his hands and seemed confused about where to put them.

"Oh no! Oh no! Oh no!" The boy moaned. Jordan stepped out from around the car.

"Hi," she said.

The boy jumped, "Oh, er...Hi."

"What are you up to?" Jordan asked. The boy looked shamefully at the three pictures in his hands. Then he looked back up at Jordan as if deciding whether or not she could be trusted. Apparently, he thought that she was trustworthy enough and confessed, "My dad is giving a presentation in a little while about the different kinds of solar eclipses, but had to run off to fix a computer problem before he could finish his poster. So he asked me if I could finish putting up these picture of different solar eclipses."

He revealed the three photos, each of which contained dark circles covering a portion of the Sun. "I was, kind of, playing on my phone while he was explaining where each picture was supposed to go and now I can't remember where to put them! He's going to be back soon and he will be so disappointed if his poster isn't finished." The boy slumped back in his chair with a look of defeat on his face.

"Well, maybe I could help." Jordan said, feeling sorry for the boy.

"Do you know anything about solar eclipses?" The boy asked suspiciously.

"I know a bit," Jordan admitted, "but I see descriptions about each kind of solar eclipse on the poster, so if we work together I'm sure we can figure it out!" The boy smiled and asked, "What's your name?"

"Jordan, what's yours?"

"Luke." So Luke and Jordan went to stand in front of the poster.

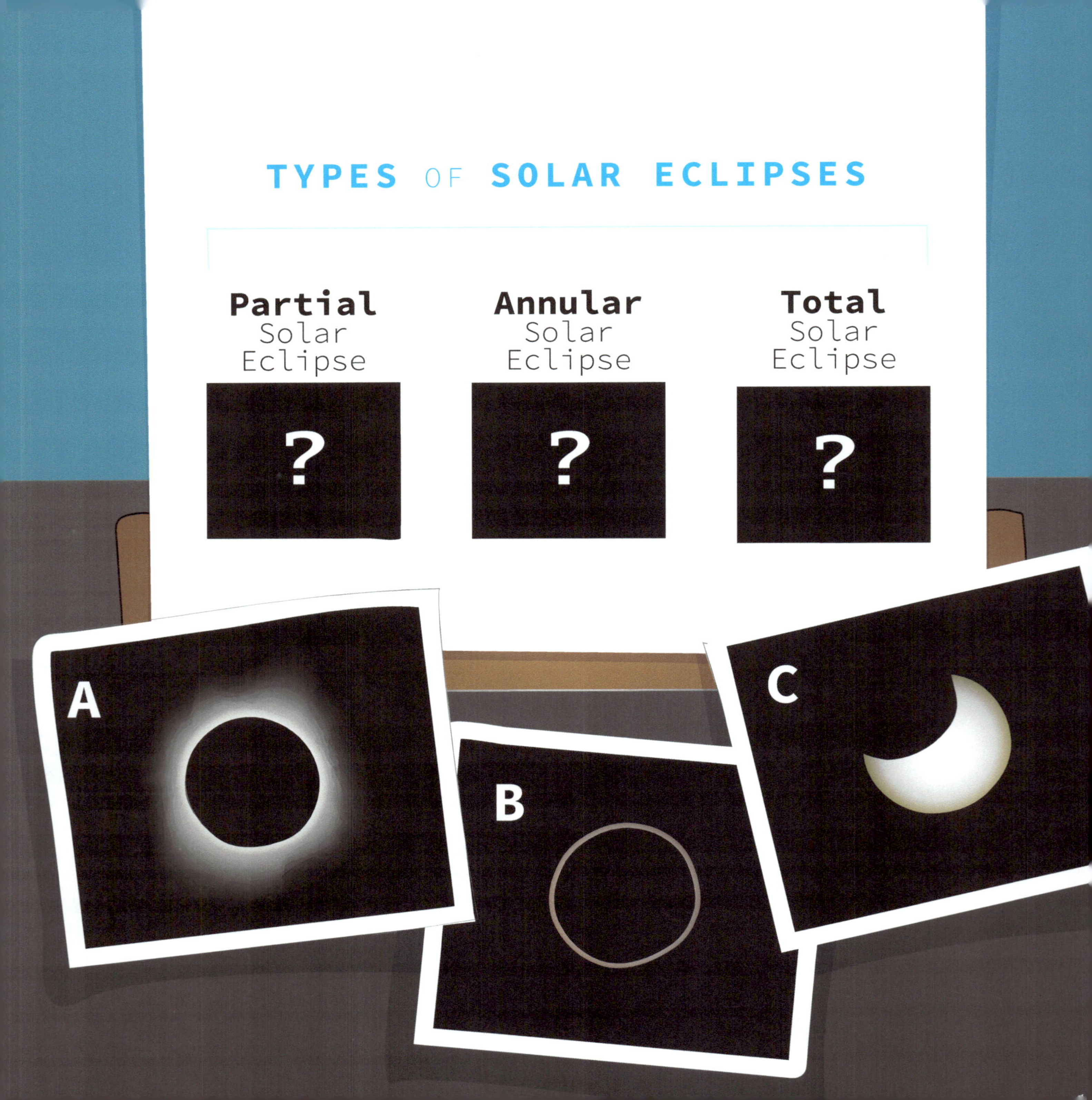

TYPES OF SOLAR ECLIPSES

Partial Solar Eclipse	**Annular** Solar Eclipse	**Total** Solar Eclipse

Jordan needs your help

Activity | Read the descriptions of each kind of solar eclipse and match each picture with its appropriate description.

1. Letter: ◯

Partial
Solar Eclipse
During a partial eclipse, the Moon only covers a portion of the Sun. It is considered partial as long as the Moon is not directly in front of the Sun.

2. Letter: ◯

Annular
Solar Eclipse
During an annular solar eclipse, the Moon is directly in front of the Sun. However, because of the elliptical orbit of the Moon about the Earth, the Moon can end up further away from us and, thus, will not cover the entire Sun.

3. Letter: ◯

Total
Solar Eclipse
During a total solar eclipse, the apparent size of the Moon is equal to or slightly greater than the apparent size of the Sun (based on orbital motions of the Moon) and the Moon is able to cover the entire face of the Sun.

Answer: 1: C, 2: B, 3: A

Bonus
Question |
What do you think is the difference between a lunar and solar eclipse?

Luke
Middle School Student

Answer: During a solar eclipse, the Moon comes between the Sun and the Earth, causing the Moon, either partially or fully, to cover the Sun. In contrast, during a lunar eclipse, the Earth comes directly between the Sun and the Moon, causing the Earth's shadow, either partially or fully, to cover the Moon.

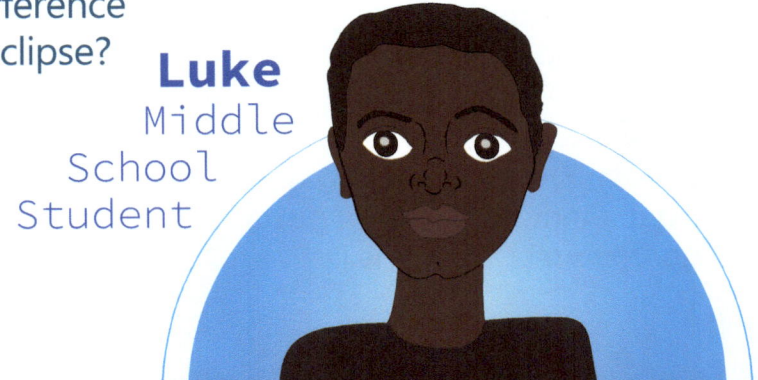

Jordan and Luke stepped back from the poster with satisfaction.

"Thank you so much, Jordan! I couldn't have done this without your help." They exchanged smiles.

All of a sudden, they heard, "Luke! I'm sorry I took so long." Luke's dad was waving at them from a tent nearby. "We gotta go! We gotta go! My presentation is in five minutes!"

"Thanks again for the help," Luke smiled. Then he picked up the poster board and sprinted away.

Jordan suddenly realized that she was starving. What time was it? Thinking back, she had only eaten a small breakfast before she and her mom had gotten in the car to come here. She started making her way back. As she walked, she looked from side to side at all of the people who were setting up

for the event. Everywhere she looked were friends, families, and strangers gathered together, laughing and talking. Even though everyone had come and claimed their own small plot of grass, just like her and her mom, the boundaries between camps were blurred by friendly faces and conversation.

She found herself comparing the sight to her cafeteria at school. Everyone hustles in to claim their small seat at the lunch table,

but at school, if you weren't friends with someone from a different table, you usually didn't go over and talk to them. That rule didn't seem to apply at this event. Here, she felt like she could walk up to any one of these groups and they would instantly greet her and be interested in what she had to say. She wondered silently why her school lunch room felt divided, but here felt united.

Out of the corner of her eye, she spotted a man who had spread a blanket on the grass and was reading a book. The title was large enough that she could read it: Myths and Misunderstandings of Eclipses. She found herself overcome with curiosity about what was inside of it. She paused and, deciding to see if people around here were really as friendly as they seemed, walked up to the man.

"Hi." She said. The man glanced up. "Hello,"

he smiled warmly and Jordan noticed he had an accent. "What can I do for you?"

"I saw you reading this book and it looked interesting. I was wondering," Jordan asked as politely as she could, "Would you tell me about it?"

"Nothing would please me more," the man said, his eyes twinkling, and he clapped his hands together revealing the entire cover.

"For thousands of years we have looked up to the skies for answers and guidance.

"Now, of course, science has helped us to understand why things happen, but imagine you are living thousands of years ago and you don't understand why things in the sky change or why the weather acts the way it does. So to help you understand these forces that are mysterious and outside of your control, you make up stories to explain why, for example, you had a drought for a few years in a row or," he smiled, "why the Sun disappeared out of the sky one day. This book is a collection of stories. Each story is a glimpse into the lives of humans who lived long ago and experienced the natural phenomenon that we now know as an eclipse." Jordan's eyes widened and her curiosity

flared even further. "What did they think was happening?"

"Well, different cultures had different explanations. But before I continue," he chuckled, "I must get something to eat. I'm afraid I've been reading here for many hours and did not realize the time! It's a good thing you came by to bring me back to reality. My name is Parmesh," the man said, reaching out his hand to shake Jordan's. "Would you like something to eat as well?"

Realizing that this must be the interesting man Hue had mentioned earlier, Jordan nodded.

They sat down with snacks and Parmesh began again.

"As you can imagine, having the Sun disappear one day would have been very disturbing for many people hundreds of years ago. The Sun and the Moon are constants in the lives of everyone and you expect them to follow a certain pattern every day. But during a solar eclipse, that pattern is disrupted. Day becomes night, if only for a few minutes. So there is a general theme in the lore behind eclipses – that is, the theme of the order in the universe being disrupted. Let's see now what we can find." Parmesh flipped through the pages of the book and they began reading.

VIKINGS » Vikings would see a pair of sky wolves chasing the Sun or the Moon. An eclipse occurs when the wolves catch the Sun or the Moon.

KOREA » It was believed that the king ordered mythical canines to capture the Sun or the Moon. When they bit either orb, you would have an eclipse.

VIETNAM » A frog or a toad would eat the Moon or the Sun during an eclipse.

BENIN & TOGO During an eclipse, it is believed that the Sun and the Moon are fighting, so everyone encourages them to stop fighting. An eclipse is often seen as a time to come together and resolve old problems.

NAVAJO » The universe is in balance, except during an eclipse when the Sun and the Moon deviate from their normal paths. An eclipse is part of nature's course, so during such an event, you take a moment to acknowledge the phenomenon and reflect on the cosmic order of things. For some, this can mean staying inside, fasting, and singing special songs. One tip from the Navajo: do not look at the Sun during an eclipse or it may affect your eyes later and make you go out of balance with the universe.

Read along with Jordan

and Parmesh about the myths of eclipses throughout the ages.

GREECE » The Ancient Greeks believed that an eclipse was the result of the wrath of the gods and was an omen of disaster and destruction.

INDIA » As the story goes, Rahu, the mythical lord, was trying to steal an elixir that would make him immortal. The Sun and the Moon, though, saw what Rahu was doing and went to tell the god, Vishnu, about Rahu's crime. Vishnu sliced off the head of Rahu before the elixir was able to get past his throat. As a result, Rahu's head became immortal, but the rest of his body died. After that, Rahu's head chased the Sun and the Moon through the sky out of hatred. An eclipse occurs when Rahu catches and swallows the Sun or the Moon. However, because Rahu has no throat, the Sun and the Moon fall out of the bottom of his head shortly after, and continue through the sky.

ENGLAND & BABYLON » The English used to believe that an eclipse was a bad omen for kings. Henry I died shortly after the 1133AD eclipse which confirmed this belief for many. In Babylon, they would put fake rulers on the throne the day of an eclipse to protect the real king.

Parmesh
Space
Historian

Jordan could hardly pull herself away from the book. Each story seemed more exciting than the last and they reminded her of the stories she and her mom would make up about constellations years ago. She suddenly had an urge to lie down in a dark grassy field and stargaze.

"It must have been so confusing for people living back then to see a solar eclipse. It's weird, I always thought space and astronomy were all about science and math and other complicated things. I didn't realize how much people have depended on the skies throughout time." Jordan felt a new appreciation for the subject.

Parmesh laughed and nodded. "Yes, many find astronomy to be intimidating. But I find that when you tell people stories about space, even real life stories about astronomers who have traveled the world to study the stars and inventions that have changed our way of life, they get more relaxed. They let themselves ask the questions they were nervous about asking before. That's why I became a space historian. I enjoy breaking that barrier and telling the human side of astronomy."

Jordan decided she had a new career goal: Space historian.

Parmesh, apparently, could read her thoughts and added, "You are a smart girl. Whether you are interested in space history or the physics that explain how the universe works, I have no doubt that you will be successful. I always tell my daughters: don't let your fear of something prevent you from following your heart."

Jordan smiled at the compliment. "Where are your daughters now?"

"At home!" Parmesh chuckled, "They did not think this event would be very exciting. I have to say I am impressed that you have decided to come to an event like this at your age. I'm sure you won't regret it."

Jordan felt a slight tinge of guilt. "Yeah, that's what my mom told me. I actually wasn't too excited to come here in the first place. I just didn't think it would be fun. But I'm having a lot more fun than I thought I would."

"Then you are all the more mature for coming into this experience with an open mind, even when you didn't want to come." Parmesh said, and then added, "You should hang out with my daughters!"

They both laughed and spent a while longer talking about famous astronomers and their expeditions across the world to uncover the mysteries of the universe. Finally, Jordan checked her phone and realized she had two missed calls from her mom, whose name was lovingly saved as, 'The Enforcer.' "I was supposed to meet my mom at our car an hour ago!" Jordan exclaimed. She thanked Parmesh for his time and his stories and started walking back toward her car.

She had only been walking for a couple of minutes when she spotted a woman painting a magnificent orb. Again, Jordan's curiosity got the best of her. She peered over the woman's shoulder and uttered a soft "wow" that was just loud enough for the women to hear.

"Oh!" The woman jumped and dropped her paint brush. "Sorry, you startled me," she gasped and laughed as she stooped down pick up her brush.

"I'm so sorry!" Jordan apologized, "I just saw your painting and wanted to get a closer look. I didn't mean to scare you."

"Ah, no worries. So you like my painting?"

She smiled and raised her eyebrows. The woman had jet black hair that fell in waves below her shoulders and a bandana keeping it away from her face. She wiped a paint-stained hand across her forehead. "Do you know what it is?"

This question took Jordan by surprise. "Well it kind of looks like the Sun, but prettier and more artistic. Like a Sun you would see in a science fiction movie."

The woman threw her head back and laughed heartily. "Oh, I love my job," she sang and picked up the painting she was working on. "It certainly is beautiful and it might please you to know that this is actually our Sun, but seen through a different point

on the electromagnetic spectrum."
Jordan furrowed her eyebrows. "The electromagnetic spectrum? Isn't the Sun yellow? Or orange or something? I've never seen it blue or green or purple."

The woman picked up a blank canvas and some colors and started painting a wave.

"Well, you're not wrong. Think of a wave that starts off long and gradually gets shorter and shorter. That is the electromagnetic spectrum. It's the range of all the possible wavelengths of light. We can only see in the visible range, which is just a small part of the wave." She drew a little box around the middle of the wave.

"So to us, because we can only see in the visible range with our eyes, the photons coming from the Sun appear yellow or maybe a little orange. But there's all of this other light," she gestured to the entire wave, "that is also being emitted by the Sun that we can't see.

"Now some animals, like bees, see in the ultraviolet spectrum. This spectrum has wavelengths that are a bit shorter than waves in the visible spectrum, so these animals may see things we can't see or in different colors." She gestured to her painting again. "This is our Sun as seen through the ultraviolet spectrum." Jordan gaped at the beauty. "Though, that's obviously not what we would normally see when we look at the Sun through a telescope."

Jordan bit her tongue, like she sometimes did when she was thinking extra hard. "How do we know, though, that the Sun looks like that through the UV spectrum if we can only see in the visible part?"

electromagnetic spectrum

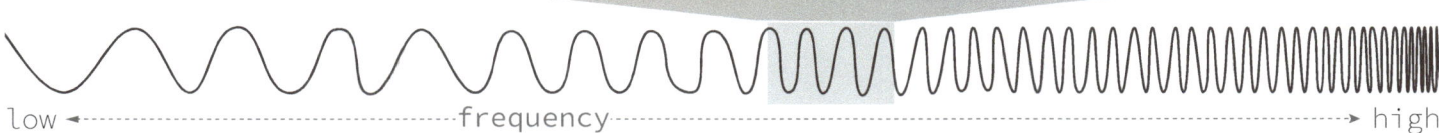

"Good question!" The woman exclaimed.

"On our own, we can't see ultraviolet light, so we use instruments to detect it for us. This is called 'remote sensing spectroscopy'. Now, even if we could see in the ultraviolet, or UV, spectrum, our atmosphere would filter out most of the UV wavelengths (thanks to our ozone layer) and most infrared wavelengths. Some wavelengths, like extreme UV, X-rays, and gamma rays, can't pass through our atmosphere at all. So we send satellites with remote sensing instruments into Earth orbit, above our atmosphere. This way, we can observe our Sun and our universe in wavelengths that we can't detect as easily here on Earth." She put down her brush and turned to Jordan.

Jordan was trying to piece together all of this new information. "But why would we want to see the Sun in different wavelengths?"

"Ah, each wavelength gives us different information about the objects we are looking at because each wavelength has slightly different properties...each one relates to different temperatures and elements. Gamma

rays and X-rays, for example, are good at telling us where areas with the hottest temperatures are located. The UV spectrum, on the other hand, can help us identify and learn more about our youngest and hottest stars that formed only a million years ago. The near infrared wavelength can penetrate clouds of dust, which helps us see farther into space."

Jordan thought she was beginning to understand. "So each wavelength kind of has its own super power, right? Because each wavelength shows us something different that we couldn't see before."

"Yes! That's a very creative analogy." The woman smiled and held out her hand. "Well, I don't know if I can leave this conversation without knowing the name of such a smart girl. I'm Charlotte, your friendly neighborhood heliophysicist." When Jordan looked confused, Charlotte quickly added, "That means that I study the Sun."

"It's a pleasure to meet you, Charlotte. I'm Jordan. And, to be honest, a lot of what you just said doesn't totally make sense to me, but thank you for the compliment."

"Oh hushpuppies!" Charlotte huffed, waving her hands dismissively. "You are inquisitive and creative. These concepts are not easy even for grown-ups to understand, let alone students, but the important thing is that you keep asking questions! Our universe is a hugely complex system and the only reason we've gotten to where we are today is because curious people, like you, kept asking questions – EVEN when others thought it was a silly question." Charlotte winked. "So keep asking! Never stop questioning the world around you."

Jordan swelled with a sense of empowerment, like the world was at her fingertips and there was all of this information just waiting for her to learn. She wanted to know it all.

Charlotte went on to explain that a total solar eclipse was the only time that the Sun's corona (the Sun's outer atmosphere) could be seen by the human eye. Jordan kept asking questions.

"So which part of the solar eclipse is the corona, again?" Jordan asked.

1 total solar eclipse

2 the sun in different wavelengths

| radio waves | micro- waves | infrared | visible | ultraviolet | x-rays | gamma rays |

Jordan needs your help

Activity | 1. Help Jordan identify the corona during a solar eclipse.

Answer: The corona is the white glow surrounding the dark Moon.

Bonus Activity | 2. Choose a wavelength and draw or paint your own Sun in that wavelength. Be as accurate or as creative as you want!

Charlotte
Heliophysicist & Painter

As Jordan waved goodbye to Charlotte, her mind was ablaze with questions about the world and the universe. All of a sudden, Jordan heard a woman exclaim, "No! Not again!" Jordan looked around and saw a woman with a floppy hat sitting at a table. She was looking desperately at her laptop screen, which was completely black.

Jordan walked over and asked the woman what was wrong. "Oh," the woman sighed, "my laptop just shut down. I was hoping to upload a short animation about the solar eclipse and a map of the path of totality, but my laptop ran out of power." The woman giggled a little and added, "And I forgot my charger." She sighed again.

"What's the path of totality?" Jordan asked.

The woman looked a little surprised and said, "Oh, well the path, um...I'm sorry, what's your name?"

"Jordan." Jordan replied.

"Ah, great name...same as my husband! I'm Laura." She waved distractedly at a fly buzzing around her head for a moment. Jordan giggled to herself as she imagined that the fly was the Moon and Laura was the Earth. Laura resumed talking as if she had never stopped.

"The path of totality describes the area on the surface of the Earth where you can see the entire eclipse. It's usually only about 100 to 200 kilometers in diameter and it makes a path across the Earth like this," Laura drew an arc in the air.

"So, if you're within the arc, the Moon will completely block out the Sun for a couple of minutes. If you're outside the path of totality, the Moon won't block the Sun out completely."

"Oh, I didn't know that." Jordan said. She glanced around briefly and saw a map sticking out of a bag near the table.

"Could you draw the path of totality on that map?" Jordan hesitated a bit. "We could even make a paper cut-out of the Moon over the Sun. And we could record a short movie using your phone... I mean, if you want." She trailed off, unsure if what she had said was just a silly idea.

Laura stared at her for a moment and Jordan started to play with her hands nervously. She wondered if she should have shared her idea or kept it to herself. But then Laura's face broke out into a grin and she clapped her hands together, shaking her head.

"And they call me a graphics animator. That's a great idea!" She exclaimed.

Jordan blushed with pride. She mused, "A graphics animator?"

"Yeah, I'm a Space Graphics Animator. I make animations of things that happen in our solar system and our universe to help people understand what's really going on out there." Laura used her hands a lot as she spoke, gesturing up to the sky. Her enthusiasm was infectious.

Jordan was instantly intrigued and made a mental note to remember the job title of 'Space Graphics Animator'. She decided she would research it more on her phone on the car ride home, along with all the other things she had learned that day.

"Ready to get started?" Laura asked, pulling out her map, some construction paper, scissors, tape, and markers.

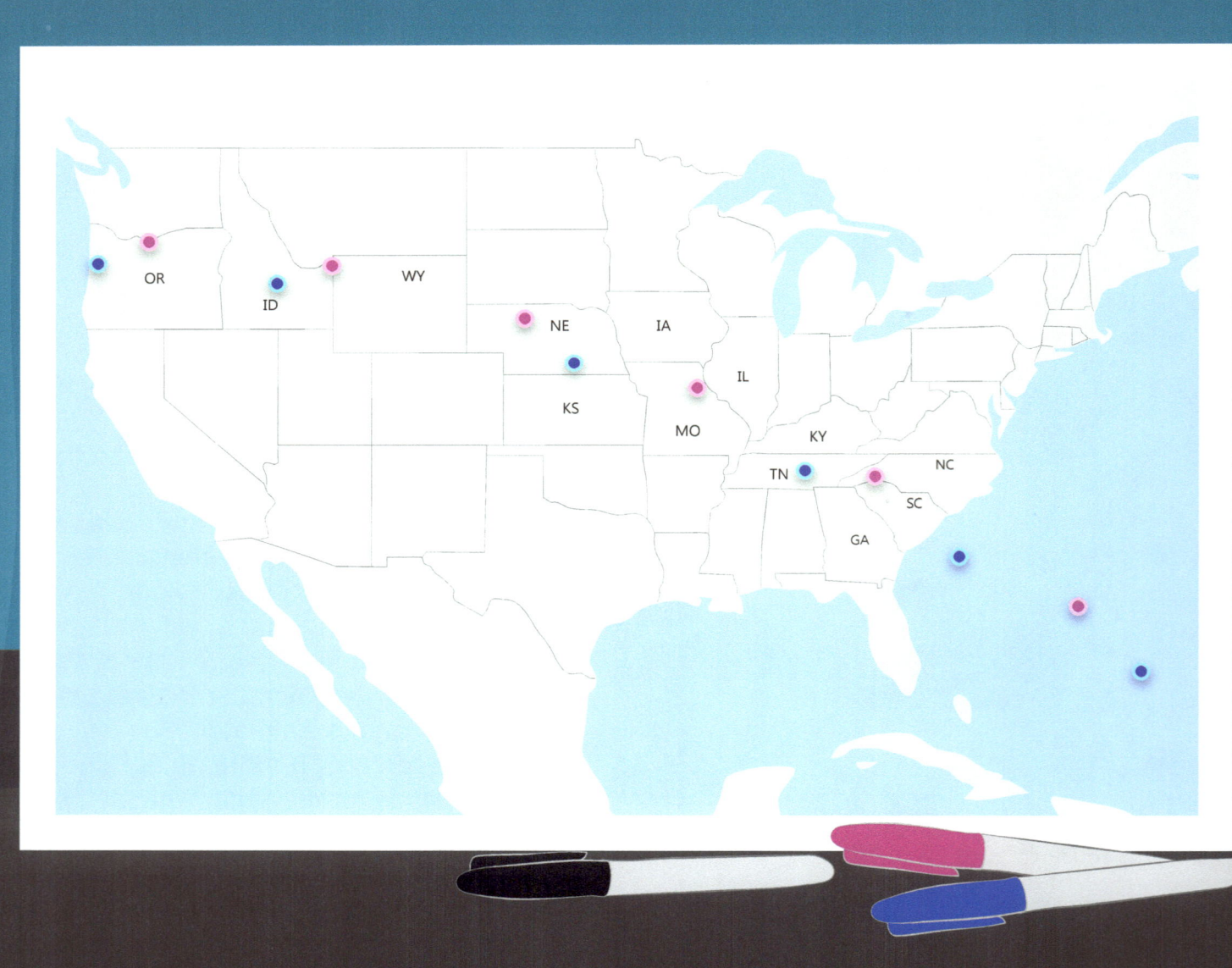

Jordan needs your help

Activity | Map the path of totality on the map below.

Laura has placed some points on the map that she knows fall within the path of totality. Follow the steps below to map the outer edges of the path of totality:

1. Connect the magenta dots with one line to create one path.
2. Connect the blue dots with another line to create another path.

To find the path of totality, draw a line down the center of both paths. This line represents the areas that will see a total solar eclipse for the longest amount of time.

The path of totality for the eclipse Jordan is going to see falls within 14 U.S. states (based on the 2017 Solar Eclipse path of totality): Oregon, Idaho, Montana, Wyoming, Nebraska, Kansas, Iowa, Missouri, Illinois, Kentucky, Tennessee, Georgia, North Carolina, and South Carolina.

Laura
Space
Graphics
Animator

When Laura and Jordan were done filming the Sun and the Moon moving across the path of totality, Laura uploaded the video using her phone. The counter showing views started to increase before Jordan's eyes and she couldn't wait to share the video with her friends at school. After watching it, maybe they would even be interested in making a few science videos with her. The possibilities seemed endless in that moment.

Jordan was having so much fun that she barely heard the announcement over the loud speaker, "10 seconds 'til first contact!" Crowds of people started to cheer and whoop and whistle. A countdown started. "Ten! Nine!..."

"It's starting!" Jordan exclaimed. She waved goodbye to Laura and ran to her mom's car where a telescope with a solar filter was set up, ready to see the Sun. Her mom was relieved to see her, but refrained from asking questions. Jordan joined in with the crowd to finish the count. "...Two! One!" The crowd chanted.

`Jordan took out her solar glasses, put them on, and looked up.`

There was the Moon, a black orb, whose side just kissed the edge of the Sun. Minute by minute the large black mass edged forward.

Jordan thought about the myths she had learned about that day. It really did look like something was eating the Sun.

"Did you know that the Moon is 400 times smaller across than the Sun?" Jordan said. The words slipped out, partly out of amazement and partly to quiz her mom. Jordan's mom looked down, surprised.

"And the only reason it covers the Sun is because the Sun is actually 400 times farther away from the Earth than the Moon is, so they seem like they're the same size." Before her mom could respond, even though she wasn't sure what to say, Jordan looked up at her and asked, "Can I look through the telescope?"

Maria smiled and stepped aside to let her daughter look. "Remember that the only reason we can look through this telescope at the Sun is because I put a solar filter on it," Maria reminded Jordan. There was the yellow-orange Sun with what looked like a small bite taken out of its side.

"Looks like the astronomers were right," someone bellowed sarcastically, "there is going to be an eclipse today!" Jordan looked at her mom and they both laughed and rolled their eyes.

phases
of a
solar
eclipse

Partial
Solar Eclipse

Baily's
Beads

Diamond
Ring

More time went by as the Moon continued to eat away at the brilliant Sun. Jordan wondered what the event would look like in the UV spectrum, recalling how beautiful Charlotte's picture had been. The Moon was now nearly covering the entire Sun. A shadow fell over the field like a blanket coming to tuck in all of the daytime creatures who were almost certainly wondering how night had come so fast. Amidst the cheers, clapping, and camera shutter clicks, Jordan noticed crickets chirping and some stars twinkling in the sky. A calm, and almost eerie, feeling that comes along with the quiet of night fell over them.

As the Moon moved to nearly cover the Sun, beads of light flared around the edges. Jordan gathered from the shouts that these were called Baily's Beads. And then, all of the beads converged into one briliant diamond-bright bead, connected by a silver ring. "There's the 'Diamond Ring'!" Someone gasped. Jordan stared silently in awe at the beatuy.

The light from the Sun was soon all but extinguished. The Moon was now fully covering the Sun and a brilliant white light glowed from behind the black Moon. The eclipse had reached totality. Everyone took off their solar glasses and looked up. Jordan had never seen anything so beautiful and radiant in her life and exclaimed, "There's the corona!"

Maria was speechless, and turned away from the dazzling sight to look at her daughter with pride. Though Jordan was growing up,

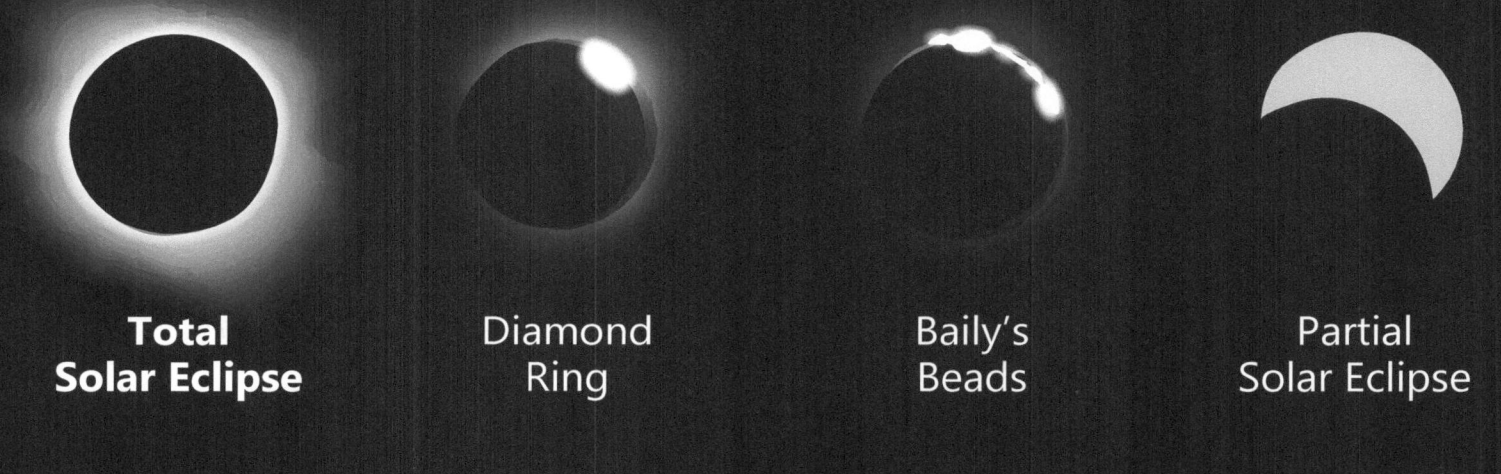

Total Solar Eclipse   Diamond Ring   Baily's Beads   Partial Solar Eclipse

the times they had spent under starry skies together all those years ago flooded back to her and she smiled and sighed a deep, happy sigh.

Totality lasted only about two and a half minutes and soon the Moon had crossed the Sun and light flooded the field again. Jordan put her glasses back on to view the end of the partial phase of the eclipse. The clicks of cameras continued vigorously for the next hour and the cheering came to a crescendo when the Moon finally kissed the Sun goodbye.

"That was amazing!" Jordan said, cheering along with the crowd. "We were so lucky to have been within the path of totality so that we could see the whole thing! How often do solar eclipses happen?"

"Well," her mom thought for a moment, "a total solar eclipse is viewable somewhere on the Earth every 18 months or so. But to see a total solar eclipse from exactly where we are again could take hundreds of years."

Jordan felt so lucky to have been able to see the total solar eclipse so close to her home. "I hope I get to see another one again." Maria smiled. "If you want to, I'm sure you will."

Jordan spotted Luke walking across the field with his dad. Charlotte and Laura were packing up not too far away and Parmesh was sitting on his picnic blanket, talking on the phone. All of her friends were getting ready to go, but Jordan didn't feel ready. She wanted to stay and learn more, ask more questions, meet more people. This day wasn't long enough! Then, she had an idea.

"Mom! I want you to come meet some friends of mine." Jordan took her mom's hand and started rounding up the friends she had made throughout the day. Soon, everyone was standing together: Hue with his camera still at his side, Olivia with her paper cut-out of the Sun, Luke, Parmesh and his book of myths, Charlotte and her paint-stained hands, and Laura in her floppy hat with the map folded in her bag. All were talking and laughing like friends, even though most of them had just met. But that wasn't surprising to Jordan. Besides a shared interest in science, there seemed to be something else special going on. Maybe there was something about the solar eclipse that reminded people that we are all in this life together, here on Earth.

"Can I ask a favor?" Jordan asked Hue. "Would you take a picture of all of us?" Hue smiled wide and swung his camera from around his neck. "It would be my pleasure."

In a few minutes, Hue had his camera set up on a tripod and joined the group. "OK, everyone. Say, 'Solar eclipse!'"

"Solar eclipse!" everyone said in unison, and then laughed. Click.

Jordan said a grateful goodbye to everyone and before she knew it, she was in the car again.

The Sun was just starting to set over the horizon. Jordan rubbed her eyes and stared out the car window and wondered if anything like this would ever happen again. Then, she smiled and remembered, if I want to do this again, I will...

And she did...

DR. JORDAN

Jordan
NASA
Astronomer

About the Story Creators

Cara Mayo, Author

Cara works as a Communications Specialist in support of a U.S. federal science agency (NOAA Fisheries Office of Science and Technology). Growing up, she was exposed to space science through her dad (an astronomer) and as a result attended many astronomical events, like solar eclipses, all over the world.

Eventually, she got involved with space science outreach which grew into a love for science communication. She went on to study environmental science in the U.S., France, and India; managed a program for a start-up; co-authored a research paper; has now written a book about space science; is currently working on her Masters in Geospatial Information Sciences and learning how to code; and who knows what next! The number of failures, mistakes, and bumps along the way are too many to list, but the journey is too exciting to stop.

Jacqui Fenner, Designer

Jacqui works as a Visual Designer in support of a U.S. federal science agency (NOAA Fisheries Office of Science and Technology). She began her career studying environmental science (fisheries science and geospatial data analysis) and eventually meandered her way to a new career at the intersection of science, communication, and visual design. Since finding her passion, she hasn't looked back.

Much to her delight, Jacqui still gets to work closely with scientists to design a range of science communication materials for both web and print. This includes websites, infographics, presentations, brochures, publications, and more. Best of all, she gets to learn and create everyday, all while contributing to the greater science community as part of a talented team of people.

Collaboration Is Key To Success // This is the philosophy that the story creators take into every project. While Cara took the lead on writing and Jacqui took the lead on design, both of them worked closely together to develop the story, characters, and layout of the book. We encourage you, dear readers, to do the same. Seek out others that inspire you and keep them close. Spend time together, work together, and challenge each other. Never forget that together we can achieve great things.

Onward, explorers! Science awaits.